Also by Balungi Francis

Unified Field Theory in a Nutshell1: The Quest for the Theory of
Everything
Fundamental Physics Research
Brief Solutions to the Big Problems in Physics, Astrophysics and
Cosmology second edition

Table of Contents

UNIFIED FIELD THEORY IN A NUTSHELL1

THE QUEST FOR THE THEORY OF EVERYTHING

BALUNGI FRANCIS

1

ALSO BY BALUNGI FRANCIS

Quantum Gravity in a Nutshell1

Fifty Formulas that Changed the World

Brief Solutions to the Big Problems in Physics, Astrophysics and cosmology

What is Real?:Space Time Singularities or Quantum Black Holes?Dark Matter or Planck Mass Particles? General Relativity or Quantum Gravity? Volume or Area Entropy Law?

Balungi's Guide to a Healthy Pregnancy

Think Physics Complete Series

Introduction

The first successful field theory was created by the Scottish Physicist James Clerk Maxwell in the mid -19th century – the electromagnetic field theory.

Then in the early 1900s, Einstein spent 30yrs of his life developing a unified field theory but all in vain.

One limitation with Einstein's work was at first his rejection, and then his ignorance of quantum theory.

Einstein was motivated by an intellectual need to unify the forces of nature. He felt very strongly that all of nature must be described by a simple force field theory from which all other forces can be deduced.

In principle, if one had the formula of a Unified field theory at hand, it would be possible to derive every other natural law from them, including all the laws of physics, chemistry, even molecular biology.

Without the use of the mathematics of the standard model and the notions of Einstein's general relativity theory, we postulate a single force field equation from which all other physics laws are derived.

A unified field theory

The building blocks of all physical laws are fields over a surface S with a defined area A.

In physics, a field is an area under the influence of some force, such as gravity or electromagnetism.

A field theory therefore refers generally to why physical phenomena happen, and how these phenomena interact with nature.

It is now clear that all forces between any two particles can easily be described by the constant interchange of quanta (force carrier or messenger particle), between them. The quanta for the electromagnetic force is a photon, the quanta for the weak force are the W and Z bosons, the quanta for the strong force is the gluon and the quanta for gravity is the graviton.

Because the vacuum is not empty but full of virtual particles popping in and out of existence in form of electromagnetic waves, the above given description of force as an exchange of quanta is replaced by a beautiful explanation in which a field is taken to be a universal sea of quantum vacuum fluctuations "the ether" on which all particles play. Therefore the force is no longer due to the exchange of quanta (as in standard model) or the curvature of space time (in general relativity) but rather the result of the quantum vacuum fluctuations.

Therefore what we observe as gravity, electromagnetism or strong force is but a manifestation of the same force (a quantum vacuum force) that connects everything

$$F = \sqrt{4\pi \frac{B}{B_o} \hbar c \rho} \qquad \qquad (1)$$

Where,

B is the minimum magnetic field limit

B_o is the maximum magnetic field limit

$\hbar$ is the reduced Planck constant

ρ is the vacuum energy density

Where c is the constant speed of light in the vacuum and ? is the coupling which determines the strength of the force. This coupling is given as

$$\alpha = \frac{B}{B_o}$$

Note: $B_o \neq B$ or $B_o \neq E/c$

Gravity

What we call space-time or "ether" are piles of surfaces each existing at a different length scales from one another. These surfaces represent the magnetic and electric fields created by a charged particle. There is a maximum magnetic field surface and the minimum magnetic field surface.

The minimum magnetic field depends on the mass of the particles involved. In gravity the first surface occurs at the Planck length and the second surface occurs at Compton wavelength as we shall see.

The measure of the magnetic field through a given surface S is called the magnetic flux. From Gauss's law, we have Maxwell's equation as,

$$\Phi_B = BS = \frac{e}{\varepsilon_0 c}$$

Where

e- is the electric charge on an electron inside the surface S

ε_0- electric constant (permitivity of free space)

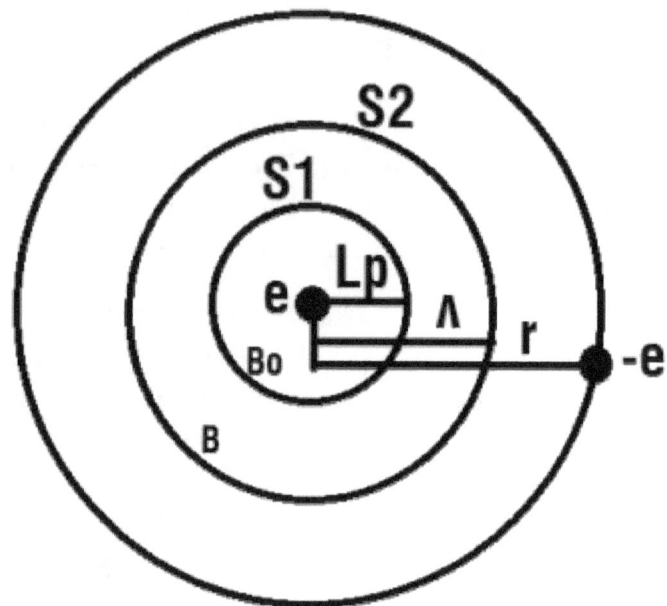

For gravity to work, the following two conditions must work;

a) The minimum magnetic field value is obtained when the radius of the surface S is the compton wavelength λ

$$S = 4\pi\lambda^2 = \frac{4\pi\hbar^2}{m^2c^2}$$

The minimum magnetic field is therefore deduced

$$B = \frac{m^2ec}{4\pi\varepsilon_0\hbar^2}$$

b) The maximum magnetic field value for gravity is obtained in the same way. This value occurs in a limit when the radius of the surface S is the Planck length l_p

$$S = 4\pi l_p^2 = \frac{4\pi\hbar G}{c^3}$$

The maximum magnetic field is therefore deduced

$$B_o = \frac{ec^2}{4\pi\varepsilon_o\hbar G}$$

Inserting into equation (1)we have

$$F = \sqrt{4\pi Gm^2\rho} \qquad\qquad (2)$$

An electron at a distance r from the electric charge causing the field feels a force F. The area of the surface at which this electron feels a force is given by $A = 4\pi r^2$. This then gives the energy density or pressure (force per unit area) as

$$\rho = \frac{F}{A} = \frac{F}{4\pi r^2}$$

Inserting in (2) we obtain a familiar law

$$F = \frac{Gm^2}{r^2}$$

We have recovered Newton's law of gravitation, practically from first principles!

These equations do not just come out by accident. It had to work, partly for dimensional reasons. The origin of gravity is a result of quantum vacuum polarization. That is the main statement, which is new and has not been made before. If true, this should have profound consequences.

Electromagnetism

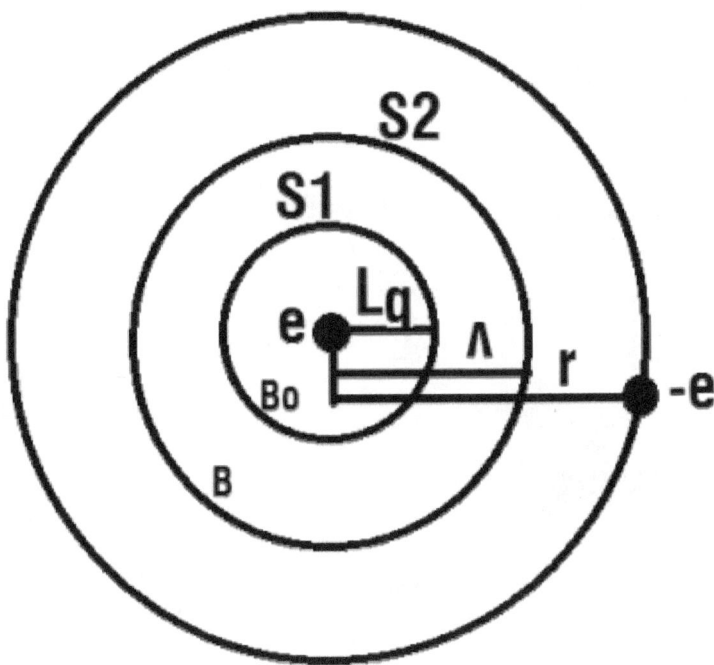

For electromagnetism to work, the following two conditions must work;

a) The minimum magnetic field value is obtained when the radius of the surface S is the compton wavelength λ

$$S = 4\pi\lambda^2 = \frac{4\pi\hbar^2}{m^2c^2}$$

The minimum magnetic field is therefore deduced

$$B = \frac{m^2ec}{4\pi\varepsilon_o\hbar^2}$$

b) The maximum magnetic field value for electromagnetism (Schwinger limit) is obtained in the same way. This value occurs in a limit when the radius of the surface S is the quantum length l_q

$$S = l_q{}^2 = \frac{\hbar e^2}{m^2 c^3 \varepsilon_0}$$

The maximum magnetic field is therefore deduced

$$B_0 = \frac{m^2 c^2}{\hbar e}$$

Inserting into equation (1) we have

$$F = \sqrt{\frac{e^2}{\varepsilon_0}}\, \rho \qquad\qquad (3)$$

An electron at a distance r from the electric charge causing the field feels a force F. The area of the surface at which this electron feels a force is given by $A = 4\pi r^2$. This then gives the energy density or pressure (force per unit area) as

$$\rho = \frac{F}{A} = \frac{F}{4\pi r^2}$$

Inserting in (3) we obtain a familiar law

$$F = \frac{e^2}{4\pi \varepsilon_0 r^2}$$

We have recovered Coulombs law of electricity, practically from first principles!

These equations do not just come out by accident. It had to work, partly for dimensional reasons. The origin of electromagnetism is a result of quantum vacuum polarization. That is the main statement, which is new and has not been made before. If true, this should have profound consequences.

Experimental proof of the Unified Theory

1. The deflection of light in the Sun's gravitational field

According to the Heisenberg Uncertainty principle, empty space isn't completely empty but consists of an indeterminate state of fluctuating fields and particles.

Even if you remove all the particles and radiation from a region of space, space still won't be empty. It will consist of virtual pairs of particles and antiparticles.

It is known that, when light propagates through an "empty" region (Vacuum), if space is perfectly empty, it should move through that space unimpeded, without bending, slowing or breaking into multiple wavelengths. Applying an external magnetic field doesn't change this, as photons, with their oscillatory electric and magnetic fields don't bend in a magnetic field. Even when your space is filled with particle antiparticle pairs, this effect doesn't change. But if you apply a **strong magnetic field** B_0 to a space filled with particle antiparticle pairs, suddenly a real, observable effect arises.

The outer 10% of a Sun consists mostly of protons, light nuclei, and electrons, which can stably exist without being crushed at the Sun's surface.

The Sun rotates frequently in excess of the 10% the speed of light, meaning that these charged particles on the outskirts of the Sun are always in motion, which necessitates the production of both electric currents and induced magnetic fields. These are the fields we should be looking for if we want to observe vacuum birefringence, and its effects on the polarization of light.

The centripetal force that keeps the particle m_1 in motion around the Sun is given as,

$$F = \frac{m_1 c^2}{R}$$

Where R is the radius of the Sun

From our unified force equation (1), we saw that for gravity the force becomes (2)

$$F = \sqrt{4\pi G m_1 m_2 \rho}$$

Here we have used m_1 and m_2 because we are dealing with interactions between different particles at the Sun's surface. However for motion to take place the following condition must be true

$$\frac{m_1 c^2}{R} = \sqrt{4\pi G m_1 m_2 \rho}$$

Since the energy density of the Sun is total energy of the Sun $(E = Mc^2)$ per unit volume $(V \sim \pi R^3)$ of the region of space in which the Sun sits,

$$\rho = \frac{Mc^2}{\pi R^3}$$

Where M is the mass of the Sun

Then inserting,

$$\frac{m_1 c^2}{R} = \sqrt{4Gm_1 m_2 \frac{Mc^2}{R^3}}$$

Squaring both sides of the equation and cancelling like terms we finally get the deflection angle of light

$$\theta = \frac{m_1}{m_2} = \frac{4GM}{Rc^2} = 1.75 \text{arcsec}$$

This value is in agreement with Einstein General Relativity theory. This is clear proof that the deflection of light is not due to curvature of space-time but is due to the vacuum polarization of space around a massive star. The deflection angle is the ratio of the mass of one particle to another (for particles near the Sun's surface).

Therefore Light from a distant star must pass through the strong magnetic field around the Sun on its way to our eyes, telescope and detectors, if the magnetized space that it passes through exhibits the expected vacuum birefringence effect, that light should all be polarized, with a common direction of polarization for all the photons.

Glossary

Absolute space and time—the Newtonian concepts of space and time, in which space is independent of the material bodies within it, and time flows at the same rate throughout the universe without regard to the locations of different observers and their experience of "now."

Acceleration—the rate at which the speed or velocity of a body changes.

Accelerating universe—the discovery in 1998, through data from very distant supernovae, that the expansion of the universe in the wake of the big bang is not slowing down, but is actually speeding up at this point in its history; groups of astronomers in California and Australia independently discovered that the light from the supernovae appears dimmer than would be expected if the universe were slowing down.

Action—the mathematical expression used to describe a physical system by requiring only the knowledge of the initial and final states of the system; the values of the physical variables at all intermediate states are determined by minimizing the action.

Anthropic principle—the idea that our existence in the universe imposes constraints on its properties; an extreme version claims that we owe our existence to this principle.

Asymptotic freedom (or safety)—a property of quantum field theory in which the strength of the coupling between elementary particles vanishes with increasing energy and/or decreasing distance, such that the elementary particles approach free particles with no external forces acting on them; moreover for decreasing energy and/or increasing distance between the particles, the strength of the particle force increases indefinitely.

Baryon—a subatomic particle composed of three quarks, such as the proton and neutron.

Big bang theory—the theory that the universe began with a violent explosion of spacetime, and that matter and energy originated from an infinitely small and dense point.

Big crunch—similar to the big bang, this idea postulates an end to the universe in a singularity.

Binary stars—a common astrophysical system in which two stars rotate around each other; also called a "double star."

Blackbody—a physical system that absorbs all radiation that hits it, and emits characteristic radiation energy depending upon temperature; the concept of blackbodies is useful, among other things, in learning the temperature of stars.

Black hole—created when a dying star collapses to a singular point, concealed by an "event horizon;" the black hole is so dense and has such strong gravity that nothing, including light, can escape it; black holes are predicted by general relativity, and though they cannot be "seen," several have been inferred from astronomical observations of binary stars and massive collapsed stars at the centers of galaxies.

Boson—a particle with integer spin, such as photons, mesons, and gravitons, which carries the forces between fermions.

Brane—shortened from "membrane," a higher-dimensional extension of a onedimensional string.

Cassini spacecraft—NASA mission to Saturn, launched in 1997, that in addition to making detailed studies of Saturn and its moons, determined a bound on the variations of Newton's gravitational constant with time.

Causality—the concept that every event has in its past events that caused it, but no event can play a role in causing events in its past.

Classical theory—a physical theory, such as Newton's gravity theory or Einstein's general relativity, that is concerned with the macroscopic universe, as opposed to theories concerning events at the submicroscopic level such as quantum mechanics and the standard model of particle physics.

Copernican revolution—the paradigm shift begun by Nicolaus Copernicus in the early sixteenth century, when he identified the sun, rather than the Earth, as the center of the known universe.

Cosmic microwave background (CMB)—the first significant evidence for the big bang theory; initially found in 1964 and studied further by NASA teams in 1989 and the early 2000s, the CMB is a smooth signature of microwaves everywhere in the sky, representing the "afterglow"of the big bang: Infrared light produced about 400,000 years after the big bang had redshifted through the stretching of spacetime during fourteen billion years of expansion to the microwave part of the electromagnetic spectrum, revealing a great deal of information about the early universe.

Cosmological constant—a mathematical term that Einstein inserted into his gravity field equations in 1917 to keep the universe static and eternal; although he later regretted this and called it his "biggest blunder," cosmologists today still use the

cosmological constant, and some equate it with the mysterious dark energy.

Coupling constant—a term that indicates the strength of an interaction between particles or fields; electric charge and Newton's gravitational constant are coupling constants.

Crystalline spheres—concentric transparent spheres in ancient Greek cosmology that held the moon, sun, planets, and stars in place and made them revolve around the Earth; they were part of the western conception of the universe until the Renaissance.

Curvature—the deviation from a Euclidean spacetime due to the warping of the geometry by massive bodies.

Dark energy—a mysterious form of energy that has been associated with negative pressure vacuum energy and Einstein's cosmological constant; it is hypothesized to explain the data on the accelerating expansion of the universe; according to the standard model, the dark energy, which is spread uniformly

throughout the universe, makes up about 70 percent of the total mass and energy content of the universe.

Dark matter—invisible, not-yet-detected, unknown particles of matter, representing about 30 percent of the total mass of matter according to the standard model; its presence is necessary if Newton's and Einstein's gravity theories are to fit data from galaxies, clusters of galaxies, and cosmology; together, dark

matter and dark energy mean that 96 percent of the matter and energy in the universe is invisible.

Deferent—in the ancient Ptolemaic concept of the universe, a large circle representing the orbit of a planet around the Earth.

Doppler principle or **Doppler effect**—the discovery by the nineteenth-century Austrian scientist Christian Doppler that when sound or light waves are moving toward an observer, the apparent frequency of the waves will be shortened, while if they are moving away from an observer, they will be lengthened; in astronomy this means that the light emitted by galaxies moving away from us is redshifted, and that from nearby galaxies moving toward us is blueshifted.

Dwarf galaxy—a small galaxy (containing several billion stars) orbiting a larger galaxy; the Milky Way has over a dozen dwarf galaxies as companions, including the Large Magellanic Cloud and Small Magellanic Cloud.

Dynamics—the physics of matter in motion.

Electromagnetism—the unified force of electricity and magnetism, discovered to be the same phenomenon by Michael Faraday and James Clerk Maxwell in the nineteenth century.

Electromagnetic radiation—a term for wave motion of electromagnetic fields which propagate with the speed of light—300,000 kilometers per second—and differ only in wavelength; this includes visible light, ultraviolet light, infrared radiation,

X-rays, gamma rays, and radio waves.

Electron—an elementary particle carrying negative charge that orbits the nucleus of an atom.

Eötvös experiments—torsion balance experiments performed by Hungarian Count Roland von Eötvös in the late nineteenth and early twentieth centuries that showed that inertial and gravitational mass were the same to one part in 1011; this was a more accurate determination of the equivalence principle than results achieved by Isaac Newton and, later, Friedrich Wilhelm Bessel.

Epicycle—in the Ptolemaic universe, a pattern of small circles traced out by a planet at the edge of its "deferent" as it orbited the Earth; this was how the Greeks accounted for the apparent retrograde motions of the planets.

Equivalence principle—the phenomenon first noted by Galileo that bodies falling in a gravitational field fall at the same rate, independent of their weight and composition; Einstein extended the principle to show that gravitation is identical (equivalent) to acceleration.

Escape velocity—the speed at which a body must travel in order to escape a strong gravitational field; rockets fired into orbits around the Earth have calculated escape velocities, as do galaxies at the periphery of galaxy clusters.

Ether (or aether)—a substance whose origins were in the Greek concept of "quintessence," the ether was the medium through which energy and matter moved, something more than a vacuum and less than air; in the late nineteenth century the Michelson-Morley experiment disproved the existence of the ether.

Euclidean geometry—plane geometry developed by the third-century bc Greek mathematician Euclid; in this geometry, parallel lines never meet.

Fermion—a particle with half-integer spin, like protons and electrons, that make up matter.

Field—a physical term describing the forces between massive bodies in gravity and electric charges in electromagnetism; Michael Faraday discovered the concept of field when studying magnetic conductors.

Field equations—differential equations describing the physical properties of interacting massive particles in gravity and electric charges in electromagnetism; Maxwell's equations for electromagnetism and Einstein's equations of gravity are prominent examples in physics.

Fifth force or **"skew" force**—a new force in MOG that has the effect of modifying gravity over limited length scales; it is carried by a particle with mass called the phion.

Fine-tuning—the unnatural cancellation of two or more large numbers involving an absurd number of decimal places, when one is attempting to explain a physical phenomenon; this signals that a true understanding of the physical phenomenon has not been achieved.

Fixed stars—an ancient Greek concept in which all the stars were static in the sky, and moved around the Earth on a distant crystalline sphere.

Frame of reference—the three spatial coordinates and one time coordinate that an observer uses to denote the position of a particle in space and time.

Galaxy—organized group of hundreds of billions of stars, such as our Milky Way.

Galaxy cluster—many galaxies held together by mutual gravity but not in as organized a fashion as stars within a single galaxy.

Galaxy rotation curve—a plot of the Doppler shift data recording the observed velocities of stars in galaxies; those stars at the periphery of giant spiral galaxies are observed to be going faster than they "should be" according to Newton's and Einstein's gravity theories.

General relativity—Einstein's revolutionary gravity theory, created in 1916 from a mathematical generalization of his theory of special relativity; it changed our concept of gravity from Newton's universal force to the warping of the geometry of spacetime in the presence of matter and energy.

Geodesic—the shortest path between two neighboring points, which is a straight line in Euclidian geometry, and a unique curved path in four-dimensional spacetime.

Globular cluster—a relatively small, dense system of up to millions of stars occurring commonly in galaxies.

Gravitational lensing—the bending of light by the curvature of spacetime; galaxies and clusters of galaxies act as lenses, distorting the images of distant bright galaxies or quasars as the light passes through or near them.

Gravitational mass—the active mass of a body that produces a gravitational force on other bodies.

Gravitational waves—ripples in the curvature of spacetime predicted by general relativity; although any accelerating body can produce gravitational radiation or waves, those that could be detected by experiments would be caused by cataclysmic cosmic events.

Graviton—the hypothetical smallest packet of gravitational energy, comparable to the photon for electromagnetic energy; the graviton has not yet been seen experimentally.

Group (in mathematics)—in abstract algebra, a set that obeys a binary operation that satisfies certain axioms; for example, the property of addition of integers makes a group; the branch of mathematics that studies groups is called group theory.

Hadron—a generic word for fermion particles that undergo strong nuclear interactions.

Hamiltonian—an alternative way of deriving the differential equations of motion for a physical system using the calculus of variations; Hamilton's principle is also called the "principle of stationary action" and was originally formulated by Sir William Rowan Hamilton for classical mechanics; the principle applies

to classical fields such as the gravitational and electromagnetic fields, and has had important applications in quantum mechanics and quantum field theory.

Homogeneous—in cosmology, when the universe appears the same to all observers, no matter where they are in the universe.

Inertia—the tendency of a body to remain in uniform motion once it is moving, and to stay at rest if it is at rest; Galileo discovered the law of inertia in the early seventeenth century.

Inertial mass—the mass of a body that resists an external force; since Newton, it has been known experimentally that inertial and gravitational mass are equal; Einstein used this equivalence of inertial and gravitational mass to postulate his equivalence principle, which was a cornerstone of his gravity theory.

Inflation theory—a theory proposed by Alan Guth and others to resolve the flatness, horizon, and homogeneity problems in the standard big bang model; the very early universe is pictured as expanding exponentially fast in a fraction of a second.

Interferometry—the use of two or more telescopes, which in combination create a receiver in effect as large as the distance between them; radio astronomy in particular makes use of interferometry.

Inverse square law—discovered by Newton, based on earlier work by Kepler, this law states that the force of gravity between two massive bodies or point particles decreases as the inverse square of the distance between them.

Isotropic—in cosmology, when the universe looks the same to an observer, no matter in which direction she looks.

Kelvin temperature scale—designed by Lord Kelvin (William Thomson) in the mid-1800s to measure very cold temperatures, its starting point is absolute zero, the coldest possible temperature in the universe, corresponding to −273.15 degrees Celsius; water's freezing point is 273.15K (0°C), while its boiling point is 373.15K (100°C).

Lagrange points—discovered by the Italian-French mathematician Joseph-Louis Lagrange, these five special points are in the vicinity of two orbiting masses where a third, smaller mass can orbit at a fixed distance from the larger masses; at the Lagrange points, the gravitational pull of the two large masses precisely equals the centripetal force required to keep the third body, such as a satellite, in a bound orbit; three of the Lagrange points are unstable, two are stable.

Lagrangian—named after Joseph-Louis Lagrange, and denoted by L, this mathematical expression summarizes the dynamical properties of a physical system; it is defined in classical mechanics as the kinetic energy T minus the potential energy V; the equations of motion of a system of particles may be derived from the Euler-Lagrange equations, a family of partial differential equations.

Light cone—a mathematical means of expressing past, present, and future space and time in terms of spacetime geometry; in four-dimensional Minkowski spacetime, the light rays emanating from or arriving at an event separate spacetime into a past cone and a future cone which meet at a point corresponding

to the event.

Lorentz transformations—

mathematical transformations from one inertial frame of reference to another such that the laws of physics remain the same; named after Hendrik Lorentz, who developed them in 1904, these transformations form the basic mathematical equations underlying special relativity.

Mercury anomaly—a phenomenon in which the perihelion of Mercury's orbit advances more rapidly than predicted by Newton's equations of gravity; when Einstein showed that his gravity theory predicted the anomalous precession, it was the first empirical evidence that general relativity might be correct.

Meson—a short-lived boson composed of a quark and an antiquark, believed to bind protons and neutrons together in the atomic nucleus.

Metric tensor—mathematical symmetric tensor coefficients that determine the infinitesimal distance between two points in spacetime; in effect the metric tensor distinguishes between Euclidean and non-Euclidean geometry.

Michelson-Morley experiment—1887 experiment by Albert Michelson and Edward Morley that proved that the ether did not exist; beams of light traveling in the same direction, and in the perpendicular direction, as the supposed ether showed no difference in speed or arrival time at their destination.

Milky Way—the spiral galaxy that contains our solar system.

Minkowski spacetime—the geometrically flat spacetime, with no gravitational effects, first described by the Swiss mathematician Hermann Minkowski; it became the setting of Einstein's theory of gravity.

MOG—my relativistic modified theory of gravitation, which generalizes Einstein's general relativity; MOG stands for "Modified Gravity."

MOND—a modification of Newtonian gravity published by Mordehai Milgrom in 1983; this is a nonrelativistic phenomenological model used to describe rotational velocity curves of galaxies; MOND stands for "Modified Newtonian

Dynamics."

Neutrino—an elementary particle with zero electric charge; very difficult to detect, it is created in radioactive decays and is able to pass through matter almost undisturbed; it is considered to have a tiny mass that has not yet been accurately measured.

Neutron—an elementary and electrically neutral particle found in the atomic nucleus, and having about the same mass as the proton.

Nuclear force—another name for the strong force that binds protons and neutrons together in the atomic nucleus.

Nucleon—a generic name for a proton or neutron within the atomic nucleus.

Neutron star—the collapsed core of a star that remains after a supernova explosion; it is extremely dense, relatively small, and composed of neutrons.

Newton's gravitational constant—the constant of proportionality, G, which occurs in the Newtonian law of gravitation, and says that the attractive force between

two bodies is proportional to the product of their masses and inversely proportional to the square of the distance between them; its numerical value is: $G = 6.67428 \pm 0.00067 \times 10{-}11$ m3 kg-1 s-2.

Nonsymmetric field theory (Einstein)—a mathematical description of the geometry of spacetime based on a metric tensor that has both a symmetric part and an antisymmetric part; Einstein used this geometry to formulate a unified field

theory of gravitation and electromagnetism.

Nonsymmetric Gravitation Theory (NGT)—my generalization of Einstein's purely gravitation theory (general relativity) that introduces the antisymmetric field as an extra component of the gravitational field; mathematically speaking, the nonsymmetric field structure is described by a non-Riemannian geometry.

Parallax—the apparent movement of a nearer object relative to a distant background when one views the object from two different positions; used with triangulation for measuring distances in astronomy.

Paradigm shift—a revolutionary change in belief, popularized by the philosopher Thomas Kuhn, in which the majority of scientists in a given field discard a traditional theory of nature in favor of a new one that passes the tests of experiment and observation; Darwin's theory of natural selection, Newton's gravity theory, and Einstein's general relativity all represented paradigm shifts.

Parsec—a unit of astronomical length equal to 3.262 light years.

Particle-wave duality—the fact that light in all parts of the electromagnetic spectrum (including radio waves, X-rays, etc., as well as visible light) sometimes acts like waves and sometimes acts like particles or photons; gravitation may be similar, manifesting as waves in spacetime or graviton particles.

Perihelion—the position in a planet's elliptical orbit when it is closest to the sun.

Perihelion advance—the movement, or changes, in the position of a planet's perihelion in successive revolutions of its orbit over time; the most dramatic perihelion advance is Mercury's, whose orbit traces a rosette pattern.

Perturbation theory—a mathematical method for finding an approximate solution to an equation that cannot be solved exactly, by expanding the solution in a series in which each successive term is smaller than the preceding one.

Phion—name given to the massive vector field in MOG; it is represented both by a boson particle, which carries the fifth force, and a field.

Photoelectric effect—the ejection of electrons from a metal by X-rays, which proved the existence of photons; Einstein's explanation of this effect in 1905 won him the Nobel Prize in 1921; separate experiments proving and demonstrating

the existence of photons were performed in 1922 by Thomas Millikan and Arthur Compton, who received the Nobel Prize for this work in 1923 and 1927, respectively.

Photon—the quantum particle that carries the energy of electromagnetic waves; the spin of the photon is 1 times Planck's constant h.

Pioneer 10 and 11 spacecraft—launched by NASA in the early 1970s to explore the outer solar system, these spacecraft show a small, anomalous acceleration as they leave the inner solar system.

Planck's constant (h)—a fundamental constant that plays a crucial role in quantum mechanics, determining the size of quantum packages of energy such as the photon; it is named after Max Planck, a founder of quantum mechanics

Principle of general covariance—Einstein's principle that the laws of physics remain the same whatever the frame of reference an observer uses to measure physical quantities.

Principle of least action—more accurately the principle of *stationary* action, this variational principle, when applied to a mechanical system or a field theory, can be used to derive the equations of motion of the system; the credit for discovering the principle is given to Pierre-Louis Moreau Maupertius but it may have been discovered independently by Leonhard Euler or Gottfried Leibniz.

Proton—an elementary particle that carries positive electrical charge and is the nucleus of a hydrogen atom.

Ptolemaic model of the universe—the predominant theory of the universe until the Renaissance, in which the Earth was the heavy center of the universe and all other heavenly bodies, including the moon, sun, planets, and stars, orbited around it; named for Claudius Ptolemy.

Quantize—to apply the principles of quantum mechanics to the behavior of matter and energy (such as the electromagnetic or gravitational field energy); breaking down a field into its smallest units or packets of energy.

Quantum field theory—the modern relativistic version of quantum mechanics used to describe the physics of elementary particles; it can also be used in nonrelativistic fieldlike systems in condensed matter physics.

Quantum gravity—attempts to unify gravity with quantum mechanics.

Quantum mechanics—the theory of the interaction between quanta (radiation) and matter; the effects of quantum mechanics become observable at the submicroscopic distance scales of atomic and particle physics, but macroscopic quantum effects can be seen in the phenomenon of quantum entanglement.

Quantum spin—the intrinsic quantum angular momentum of an elementary particle; this is in contrast to the classical orbital angular momentum of a body rotating about a point in space.

Quark—the fundamental constituent of all particles that interact through the strong nuclear force; quarks are fractionally charged, and come in several varieties; because they are confined within particles such as protons and neutrons, they cannot be detected as free particles.

Quasars—"quasi-stellar objects," the farthest distant objects that can be detected with radio and optical telescopes; they are exceedingly bright, and are believed to be young, newly forming galaxies; it was the discovery of quasars in 1960 that quashed the steady-state theory of the universe in favor of the big bang.

Quintessence—a fifth element in the ancient Greek worldview, along with earth, water, fire and air, whose purpose was to move the crystalline spheres that supported the heavenly bodies orbiting the Earth; eventually this concept became known as the "ether," which provided the *something* that bodies needed to be in contact with in order to move; although Einstein's special theory of relativity dispensed with the ether, recent explanations of the acceleration of the universe call the varying negative pressure vacuum energy "quintessence."

Redshift—a useful phenomenon based on the Doppler principle that can indicate whether and how fast bodies in the universe are receding from an observer's position on Earth; as galaxies move rapidly away from us, the frequency of the wavelength of their light is shifted toward the red end of the electromagnetic

spectrum; the amount of this shifting indicates the distance of the galaxy.

Riemann curvature tensor—a mathematical term that specifies the curvature of four-dimensional spacetime.

Riemannian geometry—a non-Euclidean geometry developed in the mid-nineteenth century by the German mathematician George Bernhard Riemann that describes curved surfaces on which parallel lines *can* converge, diverge, and even intersect, unlike Euclidean geometry; Einstein made Riemannian geometry the mathematical formalism of general relativity.

Satellite galaxy—a galaxy that orbits a host galaxy or even a cluster of galaxies.

Scalar field—a physical term that associates a value without direction to every point in space, such as temperature, density, and pressure; this is in contrast to a vector field, which has a direction in space; in Newtonian physics or in electrostatics, the potential energy is a scalar field and its gradient is the vector force field; in quantum field theory, a scalar field describes a boson particle with spin zero.

Scale invariance—distribution of objects or patterns such that the same shapes and distributions remain if one increases or decreases the size of the length scales or space in which the objects are observed; a common example of scale invariance

is fractal patterns.

Schwarzschild solution—an exact spherically symmetric static solution of Einstein's field equations in general relativity, worked out by the astronomer Karl Schwarzschild in 1916, which predicted the existence of black holes.

Self-gravitating system—a group of objects or astrophysical bodies held together by mutual gravitation, such as a cluster of galaxies; this is in contrast to a "bound system" like our solar system, in which bodies are mainly attracted to and revolve around a central mass.

Singularity—a place where the solutions of differential equations break down; a spacetime singularity is a position in space where quantities used to determine the gravitational field become infinite; such quantities include the curvature of spacetime and the density of matter.

Spacetime—in relativity theory, a combination of the three dimensions of space with time into a four-dimensional geometry; first introduced into relativity by Hermann Minkowski in 1908.

Special theory of relativity—Einstein's initial theory of relativity, published in 1905, in which he explored the "special" case of transforming the laws of physics from one uniformly moving frame of reference to another; the equations

of special relativity revealed that the speed of light is a constant, that objects appear contracted in the direction of motion when moving at close to the speed of light, and that $E = mc2$, or energy is equal to mass times the speed of light squared.

Spin—see quantum spin.

String theory—a theory based on the idea that the smallest units of matter are not point particles but vibrating strings; a popular research pursuit in physics for two decades, string theory has some attractive mathematical features, but has yet to make a testable prediction.

Strong force—see nuclear force.

Supernova—spectacular, brilliant death of a star by explosion and the release of heavy elements into space; supernovae type 1a are assumed to have the same intrinsic brightness and are therefore used as standard candles in estimating cosmic distances.

Supersymmetry—a theory developed in the 1970s which, proponents claim, describes the most fundamental spacetime symmetry of particle physics: For every boson particle there is a supersymmetric fermion partner, and for every fermion there exists a supersymmetric boson partner; to date, no supersymmetric particle partner has been detected.

Tully-Fisher law—a relation stating that the asymptotically flat rotational velocity of a star in a galaxy, raised to the fourth power, is proportional to the mass or luminosity of the galaxy.

Unified theory (or unified field theory)—a theory that unites the forces of nature; in Einstein's day those forces consisted of electromagnetism and gravity; today the weak and strong nuclear forces must also be taken into account, and perhaps someday MOG's fifth force or skew force will be included; no one has yet discovered a successful unified theory.

Vacuum—in quantum mechanics, the lowest energy state, which corresponds to the vacuum state of particle physics; the vacuum in modern quantum field theory is the state of perfect balance of creation and annihilation of particles and antiparticles.

Variable Speed of Light (VSL) cosmology—an alternative to inflation theory, in which the speed of light was much faster at the beginning of the universe than it is today; like inflation, this theory solves the horizon and flatness problems in the very early universe in the standard big bang model.

Vector field—a physical value that assigns a field with the position and direction of a vector in space; it describes the force field of gravity or the electric and magnetic force fields in James Clerk Maxwell's field equations.

Virial theorem—a means of estimating the average speed of galaxies within galaxy clusters from their estimated average kinetic and potential energies.

Vulcan—a hypothetical planet predicted by the nineteenth-century astronomer Urbain Jean Joseph Le Verrier to be the closest orbiting planet to the sun; the presence of Vulcan would explain the anomalous precession of the perihelion of Mercury's orbit; Einstein later explained the anomalous precession in general relativity by gravity alone.

Weak force—one of the four fundamental forces of nature, associated with radioactivity such as beta decay in subatomic physics; it is much weaker than the strong nuclear force but still much stronger than gravity.

X-ray clusters—galaxy clusters with large amounts of extremely hot gas within them that emit X-rays; in such clusters, this hot gas represents at least twice the mass of the luminous stars.

Bibliography

Misner, C.W., Thorne, K.S., and Wheeler, J.A. (1973) *Gravitation*, Freeman, San Francisco, p. 5.

Wilson, H.A. (1921) An electromagnetic theory of gravitation, *Phys. Rev.* **17**, 54-59.

Dicke, R.H. (1957) Gravitation without a principle of equivalence, *Rev. Mod. Phys.* **29**, 363-376. See also Dicke, R.H. (1961) Mach's principle and equivalence, in C. Møller (ed.), *Proc. of the Intern'l School of Physics "Enrico Fermi" Course XX, Evidence for Gravitational Theories*, Academic Press, New York, pp.1-49.

Lightman, A.P., and Lee, D.L. (1973) Restricted proof that the weak equivalence principle implies the Einstein equivalence principle, *Phys. Rev. D* **8**, 364-376.

Will, C.M. (1974) Gravitational red-shift measurements as tests of nonmetric theories of gravity, *Phys. Rev. D* **10**, 2330-2337.

Haugan, M.P., and Will, C.M. (1977) Princip les of equivalence, Eötvös experiments, and gravitational red-shift experiments: The free fall of electromagnetic systems to post -post-Coulombian order, *Phys. Rev. D* **15**, 2711-2720.

Volkov, A.M., Izmest'ev, A.A., and Skrotskii, G.V. (1971) The propagation of electromagnetic waves in a Riemannian space, *Sov. Phys. JETP* **32**, 686-689.

Heitler, W. (1954) *The Quantum Theory of Radiation*, 3rd ed., Oxford University Press, London, p.

113.

Alpher, R.A. (Jan.-Feb. 1973) Large numbers, cosmology, and Gamow, *Am. Sci.* **61**, 52-58.

Harrison, E.R. (Dec. 1972) The cosmic numbers, *Phys. Today* **25**, 30-34.

Webb, J.K., Flambaum, V.V., Churchill, C.W., Drinkwater, M.J., and Barrow, J.D. (1999) Search for time variation of the fine structure constant, *Phys. Rev. Lett.* **82**, 884-887.

Brault, J.W. (1963) Gravitational red shift of solar lines, *Bull. Amer. Phys. Soc.* **8**, 28.

Pound, R.V., and Rebka, G.A. (1960) Apparent weight of photons, *Phys. Rev. Lett.* **4**, 337-341.

Pound, R.V., and Snider, J.L. (1965) Effect of gravity on nuclear resonance, *Phys. Rev. Lett.* **13**, 539-540.

Goldstein, H. (1957) *Classical Mechanics*, Addison-Wesley, Reading MA, pp. 206-207.

Mizobuchi, Y. (1985) New theory of space-time and gravitation - Yilmaz's approach, *Hadronic Jour.* **8**, 193-219.

Alley, C.O. (1995) The Yilmaz theory of gravity and its compatibility with quantum theory, in D.M. Greenberger and A. Zeilinger (eds.), *Fundamental Problems in Quantum Theory: A Conference Held in Honor of Professor John A. Wheeler*, Vol. 755 of the Annals of the New York Academy of Sciences, New York, pp. 464-475.

Schilling, G.(1999) Watching the universe's second biggest bang, *Science* **283**, 2003-2004.

Robertson, S.L. (1999) Bigger bursts from merging neutron stars, *Astrophys. Jour.* **517**, L117-L119.

Hughes, V.W., Robinson, H.G. and Beltran-Lopez, V. (1960) Upper limit for the anisotropy of inertial mass from nuclear resonance experiments, *Phys. Rev. Lett.* **4**, 342-344.

Drever, R.W.P. (1961) A search for anisotropy of inertial mass using a free precession technique,

Phil. Mag. **6**, 683-687.

Collela, R. Overhauser, A.W., and Werner, S.A. (1975) Observation of gravitationally induced quantum mechanics, *Phys. Rev Lett.* **34**, 1472-1474.

Puthoff, H.E. (1996) SETI, the velocity-of-light limitation, and the Alcubierre warp drive: an integrating overview, *Physics Essays* **9**, 156-158.

Atkinson, R. d'E. (1962) General relativity in Euclidean terms, *Proc. Roy. Soc.* **272**, 60-7

Barton, G.; Scharnhorst, K. (1993). "QED Between Parallel Mirrors: Light Signals Faster Than c, or Amplified by the Vacuum". Journal of Physics A: Mathematical and General. **26** (8): 2037–2046. *Bibcode[1]:1993JPhA...26.2037B[2]. doi[3]:10.1088/0305-4470/26/8/024[4]. ISSN[5] 0305-4470[6].*

1. *https://en.wikipedia.org/wiki/Bibcode*

2. *https://ui.adsabs.harvard.edu/abs/1993JPhA...26.2037B*

3. *https://en.wikipedia.org/wiki/Digital_object_identifier*

4. *https://doi.org/10.1088%2F0305-4470%2F26%2F8%2F024*

5. *https://en.wikipedia.org/wiki/International_Standard_Serial_Number*

6. *https://www.worldcat.org/issn/0305-4470*

Beiser, A. (2003). *Concepts of Modern Physics*[7] (6th ed.). Boston: McGraw-Hill. *ISBN*[8] *978-0072448481*[9]. *LCCN*[10] *2001044743*[11]. *OCLC*[12] *48965418*[13].

Bordag, M; Klimchitskaya, G. L.; Mohideen, U.; Mostepanenko, V. M. (2009). *Advances in the Casimir Effect*[14]. Oxford: Oxford University Press. *ISBN*[15] *978-0-19-923874-3*[16]. *LCCN*[17] *2009279136*[18]. *OCLC*[19] *319209483*[20].

Boyer, T. H. (1970). "Quantum Zero-Point Energy and Long-Range Forces". Annals of Physics. 56 (2): 474–503.

---

7. *https://books.google.com/*

  books?id=34l9CgAAQBAJ&printsec=frontcover#v_43ec3e5dee6e706af7766fffea512721_onepage_6

  cff047854f19ac2aa52aac51bf3af4a_q_6cff047854f19ac2aa52aac51bf3af4a_f_43ec3e5dee6e706af7

  766fffea512721_false

8. *https://en.wikipedia.org/wiki/International_Standard_Book_Number*

9. *https://en.wikipedia.org/wiki/Special:BookSources/978-0072448481*

10. *https://en.wikipedia.org/wiki/Library_of_Congress_Control_Number*

11. *https://lccn.loc.gov/2001044743*

12. *https://en.wikipedia.org/wiki/OCLC*

13. *https://www.worldcat.org/oclc/48965418*

14. *https://books.google.com/*

  books?id=CqE1f_s5PgYC&printsec=frontcover#v_43ec3e5dee6e706af7766fffea512721_onepage_6c

  ff047854f19ac2aa52aac51bf3af4a_q_6cff047854f19ac2aa52aac51bf3af4a_f_43ec3e5dee6e706af7

  766fffea512721_false

15. *https://en.wikipedia.org/wiki/International_Standard_Book_Number*

16. *https://en.wikipedia.org/wiki/Special:BookSources/978-0-19-923874-3*

17. *https://en.wikipedia.org/wiki/Library_of_Congress_Control_Number*

18. *https://lccn.loc.gov/2009279136*

19. *https://en.wikipedia.org/wiki/OCLC*

20. *https://www.worldcat.org/oclc/319209483*

42 BALUNGI FRANCIS

$Bibcode^{21}$:*1970AnPhy..56..474B^{22}.doi^{23}:10.1016/ 0003-4916(70)90027-8^{24}. ISSN^{25} 0003-4916^{26}. OCLC^{27} 4648258537^{28}.*

Carroll, S. M.^{29}; Field, G. B. (1997). *"Is There Evidence for Cosmic Anisotropy in the Polarization of Distant Radio Sources?"^{30} (PDF).* Physical Review Letters. 79 (13): 2394–2397.*arXiv^{31}:astro-ph/ 9704263^{32}.*

Conlon, T. E. (2011). *Thinking About Nothing : Otto Von Guericke and The Magdeburg Experiments on the Vacuum^{33}.* San Francisco: Saint Austin Press. *ISBN^{34} 978-1-4478-3916-3^{35}. OCLC^{36} 840927124^{37}.*

---

21. https://en.wikipedia.org/wiki/Bibcode

22. https://ui.adsabs.harvard.edu/abs/1970AnPhy..56..474B

23. https://en.wikipedia.org/wiki/Digital_object_identifier

24. https://doi.org/10.1016%2F0003-4916%2870%2990027-8

25. https://en.wikipedia.org/wiki/International_Standard_Serial_Number

26. https://www.worldcat.org/issn/0003-4916

27. https://en.wikipedia.org/wiki/OCLC

28. https://www.worldcat.org/oclc/4648258537

29. https://en.wikipedia.org/wiki/Sean_M._Carroll

30. http://cds.cern.ch/record/324872/files/9704263.pdf

31. https://en.wikipedia.org/wiki/ArXiv

32. https://arxiv.org/abs/astro-ph/9704263

33. https://books.google.com/

    books?id=rL9jAwAAQBAJ&printsec=frontcover#v_43ec3e5dee6e706af7766fffea512721_onepage_6

    cff047854f19ac2aa52aac51bf3af4a_q_6cff047854f19ac2aa52aac51bf3af4a_f_43ec3e5dee6e706af7

    766fffea512721_false

34. https://en.wikipedia.org/wiki/International_Standard_Book_Number

35. https://en.wikipedia.org/wiki/Special:BookSources/978-1-4478-3916-3

36. https://en.wikipedia.org/wiki/OCLC

37. https://www.worldcat.org/oclc/840927124

Davies, P. C. W.[38] (1985). Superforce: The Search for a Grand Unified Theory of Nature[39]. New York: Simon and Schuster. ISBN[40] 978-0-671-47685-4[41]. LCCN[42] 84005473[43]. OCLC[44] 12397205[45].

Dunne, G. V. (2012). "The Heisenberg-Euler Effective Action: 75 years on". International Journal of Modern Physics A. 27 (15): 1260004. arXiv[46]:1202.1557[47].   Bibcode[48]:2012IJMPA..2760004D[49]. doi[50]:10.1142/S0217751X12600044[51]. ISSN[52] 0217-751X[53].

Einstein, A.[54] (1995). Klein, Martin J.; Kox, A. J.; Renn, Jürgen; Schulmann, Robert (eds.). The Collected Papers of Albert Einstein Vol. 4 The Swiss Years: Writings, 1912–1914[55]. Princeton: Princeton University Press. ISBN[56] 978-0-691-03705-9[57]. OCLC[58] 929349643[59].

---

38. https://en.wikipedia.org/wiki/Paul_Davies

39. https://archive.org/details/superforcesearch00davi

40. https://en.wikipedia.org/wiki/International_Standard_Book_Number

41. https://en.wikipedia.org/wiki/Special:BookSources/978-0-671-47685-4

42. https://en.wikipedia.org/wiki/Library_of_Congress_Control_Number

43. https://lccn.loc.gov/84005473

44. https://en.wikipedia.org/wiki/OCLC

45. https://www.worldcat.org/oclc/12397205

46. https://en.wikipedia.org/wiki/ArXiv

47. https://arxiv.org/abs/1202.1557

48. https://en.wikipedia.org/wiki/Bibcode

49. https://ui.adsabs.harvard.edu/abs/2012IJMPA..2760004D

50. https://en.wikipedia.org/wiki/Digital_object_identifier

51. https://doi.org/10.1142%2FS0217751X12600044

52. https://en.wikipedia.org/wiki/International_Standard_Serial_Number

53. https://www.worldcat.org/issn/0217-751X

54. https://en.wikipedia.org/wiki/Albert_Einstein

55. http://einsteinpapers.press.princeton.edu/vol4-doc/

56. https://en.wikipedia.org/wiki/International_Standard_Book_Number

57. https://en.wikipedia.org/wiki/Special:BookSources/978-0-691-03705-9

Greiner, W.; Müller, B.; Rafelski, J. (2012). *Quantum Electrodynamics of Strong Fields: With an Introduction into Modern Relativistic Quantum Mechanics*[60]. Springer. *doi*[61]:*10.1007/978-3-642-82272-8*[62]. *ISBN*[63] *978-0-387-13404-8*[64]. *LCCN*[65] *84026824*[66]. *OCLC*[67] *317097176*[68].

Haisch, B.; Rueda, A.; Puthoff, H. E. (1994). *"Inertia as a Zero-Point-Field Lorentz Force"*[69] *(PDF)*. Physical Review A. 49 (2): 678–694. *Bibcode*[70]:*1994PhRvA..49..678H*[71]. *doi*[72]:*10.1103/ PhysRevA.49.678*[73]. *PMID*[74] *9910287*[75].

Heisenberg, W.; Euler, H. (1936). "Folgerungen aus der Diracschen Theorie des Positrons". Zeitschrift für Physik. 98 (11–12): 714–732.

58. *https://en.wikipedia.org/wiki/OCLC*

59. *https://www.worldcat.org/oclc/929349643*

60. *https://books.google.com/*

   books?id=Wh3-CAAAQBAJ&printsec=frontcover#v_43ec3e5dee6e706af7766fffea512721_onepage

   _6cff047854f19ac2aa52aac51bf3af4a_q_6cff047854f19ac2aa52aac51bf3af4a_f_43ec3e5dee6e706

   af7766fffea512721_false

61. *https://en.wikipedia.org/wiki/Digital_object_identifier*

62. *https://doi.org/10.1007%2F978-3-642-82272-8*

63. *https://en.wikipedia.org/wiki/International_Standard_Book_Number*

64. *https://en.wikipedia.org/wiki/Special:BookSources/978-0-387-13404-8*

65. *https://en.wikipedia.org/wiki/Library_of_Congress_Control_Number*

66. *https://lccn.loc.gov/84026824*

67. *https://en.wikipedia.org/wiki/OCLC*

68. *https://www.worldcat.org/oclc/317097176*

69. *http://www.zpower.com/sp/documents/ZPEPaper_InertiaAsAZeroPointFieldLorenzForce.pdf*

70. *https://en.wikipedia.org/wiki/Bibcode*

71. *https://ui.adsabs.harvard.edu/abs/1994PhRvA..49..678H*

72. *https://en.wikipedia.org/wiki/Digital_object_identifier*

73. *https://doi.org/10.1103%2FPhysRevA.49.678*

74. *https://en.wikipedia.org/wiki/PubMed_Identifier*

75. *https://pubmed.ncbi.nlm.nih.gov/9910287*

arXiv[76]:*physics/0605038*[77].   *Bibcode*[78]:*1936ZPhy...98..714H*[79]. *doi*[80]:*10.1007/BF01343663*[81]. *ISSN*[82] *1434-6001*[83].

Heitler, W.[84] (1984). *The Quantum Theory of Radiation*[85] (1954 reprint 3rd ed.). New York: Dover Publications. *ISBN*[86] *978-0486645582*[87]. *LCCN*[88] *83005201*[89]. *OCLC*[90] *924845769*[91].

Heyl, J. S.; Shaviv, N. J. (2000). "Polarization evolution in strong magnetic fields". Monthly Notices of the Royal Astronomical Society. 311   (3):   555–564.   *arXiv*[92]:*astro-ph/9909339*[93].

76. *https://en.wikipedia.org/wiki/ArXiv*

77. *https://arxiv.org/abs/physics/0605038*

78. *https://en.wikipedia.org/wiki/Bibcode*

79. *https://ui.adsabs.harvard.edu/abs/1936ZPhy...98..714H*

80. *https://en.wikipedia.org/wiki/Digital_object_identifier*

81. *https://doi.org/10.1007%2FBF01343663*

82. *https://en.wikipedia.org/wiki/International_Standard_Serial_Number*

83. *https://www.worldcat.org/issn/1434-6001*

84. *https://en.wikipedia.org/wiki/Walter_Heitler*

85. *https://books.google.com/*
   books?id=L7w7UpecbKYC&printsec=frontcover#v_43ec3e5dee6e706af7766fffea512721_onepage_6
   cff047854f19ac2aa52aac51bf3af4a_q_6cff047854f19ac2aa52aac51bf3af4a_f_43ec3e5dee6e706af7
   766fffea512721_false

86. *https://en.wikipedia.org/wiki/International_Standard_Book_Number*

87. *https://en.wikipedia.org/wiki/Special:BookSources/978-0486645582*

88. *https://en.wikipedia.org/wiki/Library_of_Congress_Control_Number*

89. *https://lccn.loc.gov/83005201*

90. *https://en.wikipedia.org/wiki/OCLC*

91. *https://www.worldcat.org/oclc/924845769*

92. *https://en.wikipedia.org/wiki/ArXiv*

93. *https://arxiv.org/abs/astro-ph/9909339*

Bibcode[94]:2000MNRAS.311..555H[95].   *doi[96]:10.1046/*
j.1365-8711.2000.03076.x[97]. ISSN[98] 0035-8711[99].

Itzykson, C.; Zuber, J.-B. (1980). *Quantum Field Theory[100]* (2005 ed.).
Mineola, New York: Dover Publications. *ISBN[101] 978-0486445687[102].*
LCCN[103] 2005053026[104]. OCLC[105] 61200849[106].

Kostelecký, V. Alan; Mewes, M. (2009). "Electrodynamics with
Lorentz-violating operators of arbitrary dimension". Physical Review D.
80   (1):   015020.   *arXiv[107]:0905.0031[108].*
Bibcode[109]:2009PhRvD..80a5020K[110].   *doi[111]:10.1103/*
PhysRevD.80.015020[112]. ISSN[113] 1550-7998[114].

94. https://en.wikipedia.org/wiki/Bibcode

95. https://ui.adsabs.harvard.edu/abs/2000MNRAS.311..555H

96. https://en.wikipedia.org/wiki/Digital_object_identifier

97. https://doi.org/10.1046%2Fj.1365-8711.2000.03076.x

98. https://en.wikipedia.org/wiki/International_Standard_Serial_Number

99. https://www.worldcat.org/issn/0035-8711

100. https://books.google.com/

    books?id=4MwsAwAAQBAJ&printsec=frontcover#v_43ec3e5dee6e706af7766fffea512721_onepage
    _6cff047854f19ac2aa52aac51bf3af4a_q_6cff047854f19ac2aa52aac51bf3af4a_f_43ec3e5dee6e706
    af7766fffea512721_false

101. https://en.wikipedia.org/wiki/International_Standard_Book_Number

102. https://en.wikipedia.org/wiki/Special:BookSources/978-0486445687

103. https://en.wikipedia.org/wiki/Library_of_Congress_Control_Number

104. https://lccn.loc.gov/2005053026

105. https://en.wikipedia.org/wiki/OCLC

106. https://www.worldcat.org/oclc/61200849

107. https://en.wikipedia.org/wiki/ArXiv

108. https://arxiv.org/abs/0905.0031

109. https://en.wikipedia.org/wiki/Bibcode

110. https://ui.adsabs.harvard.edu/abs/2009PhRvD..80a5020K

111. https://en.wikipedia.org/wiki/Digital_object_identifier

Kostelecký, V. Alan; Mewes, M. (2013). "Constraints on Relativity Violations from Gamma-Ray Bursts". Physical Review Letters. 110 (20): 201601. *arXiv*[115]:*1301.5367*[116]. *Bibcode*[117]:*2013PhRvL.110t1601K*[118]. *doi*[119]:*10.1103/PhysRevLett.110.201601*[120]. *ISSN*[121] *0031-9007*[122]. *PMID*[123] *25167393*[124].

Kragh, H. (2012). "Preludes to Dark Energy: Zero-Point Energy and Vacuum Speculations". Archive for History of Exact Sciences. 66 (3): 199–240. *arXiv*[125]:*1111.4623*[126]. *doi*[127]:*10.1007/s00407-011-0092-3*[128]. *ISSN*[129] *0003-9519*[130].

Kragh, H. S.; Overduin, J. M. (2014). *The Weight of the Vacuum : A Scientific History of Dark Energy*[131]. New York: Springer. *ISBN*[132]

112. https://doi.org/10.1103%2FPhysRevD.80.015020

113. https://en.wikipedia.org/wiki/International_Standard_Serial_Number

114. https://www.worldcat.org/issn/1550-7998

115. https://en.wikipedia.org/wiki/ArXiv

116. https://arxiv.org/abs/1301.5367

117. https://en.wikipedia.org/wiki/Bibcode

118. https://ui.adsabs.harvard.edu/abs/2013PhRvL.110t1601K

119. https://en.wikipedia.org/wiki/Digital_object_identifier

120. https://doi.org/10.1103%2FPhysRevLett.110.201601

121. https://en.wikipedia.org/wiki/International_Standard_Serial_Number

122. https://www.worldcat.org/issn/0031-9007

123. https://en.wikipedia.org/wiki/PubMed_Identifier

124. https://pubmed.ncbi.nlm.nih.gov/25167393

125. https://en.wikipedia.org/wiki/ArXiv

126. https://arxiv.org/abs/1111.4623

127. https://en.wikipedia.org/wiki/Digital_object_identifier

128. https://doi.org/10.1007%2Fs00407-011-0092-3

129. https://en.wikipedia.org/wiki/International_Standard_Serial_Number

130. https://www.worldcat.org/issn/0003-9519

*978-3-642-55089-8[133].   LCCN[134]   2014938218[135].   OCLC[136]
884863929[137].*

Kuhn, T.[138] (1978). *Black-Body Theory and the Quantum Discontinuity,
1894-1912[139].* New York: Oxford University Press. *ISBN[140]
978-0-19-502383-1[141]. LCCN[142] 77019022[143]. OCLC[144] 803538583[145].*

Lahteenmaki, P.; Paraoanu, G. S.; Hassel, J.; Hakonen, P. J. (2013).
"Dynamical Casimir Effect in a Josephson Metamaterial". Proceedings
of the National Academy of Sciences. 110 (11): 4234–4238.

131. *https://books.google.com/*

    books?id=sroqBAAAQBAJ&printsec=frontcover#v_43ec3e5dee6e706af7766fffea512721_onepage_6

    cff047854f19ac2aa52aac51bf3af4a_q_6cff047854f19ac2aa52aac51bf3af4a_f_43ec3e5dee6e706af7

    766fffea512721_false

132. *https://en.wikipedia.org/wiki/International_Standard_Book_Number*

133. *https://en.wikipedia.org/wiki/Special:BookSources/978-3-642-55089-8*

134. *https://en.wikipedia.org/wiki/Library_of_Congress_Control_Number*

135. *https://lccn.loc.gov/2014938218*

136. *https://en.wikipedia.org/wiki/OCLC*

137. *https://www.worldcat.org/oclc/884863929*

138. *https://en.wikipedia.org/wiki/Thomas_Kuhn*

139. *https://en.wikipedia.org/wiki/Black-Body_Theory_and_the_Quantum_Discontinuity,_1894-1912*

140. *https://en.wikipedia.org/wiki/International_Standard_Book_Number*

141. *https://en.wikipedia.org/wiki/Special:BookSources/978-0-19-502383-1*

142. *https://en.wikipedia.org/wiki/Library_of_Congress_Control_Number*

143. *https://lccn.loc.gov/77019022*

144. *https://en.wikipedia.org/wiki/OCLC*

145. *https://www.worldcat.org/oclc/803538583*

arXiv[146]:*1111.5608*[147].   *Bibcode*[148]:*2013PNAS..110.4234L*[149]. *doi*[150]:*10.1073/pnas.1212705110*[151]. *ISSN*[152] *0027-8424*[153].

Leuchs, G.; Sánchez-Soto, L. L. (2013). "A Sum Rule For Charged Elementary Particles". The European Physical Journal D. 67 (3): 57. *arXiv*[154]:*1301.3923*[155].   *Bibcode*[156]:*2013EPJD...67...57L*[157]. *doi*[158]:*10.1140/epjd/e2013-30577-8*[159]. *ISSN*[160] *1434-6060*[161].

146. https://en.wikipedia.org/wiki/ArXiv

147. https://arxiv.org/abs/1111.5608

148. https://en.wikipedia.org/wiki/Bibcode

149. https://ui.adsabs.harvard.edu/abs/2013PNAS..110.4234L

150. https://en.wikipedia.org/wiki/Digital_object_identifier

151. https://doi.org/10.1073%2Fpnas.1212705110

152. https://en.wikipedia.org/wiki/International_Standard_Serial_Number

153. https://www.worldcat.org/issn/0027-8424

154. https://en.wikipedia.org/wiki/ArXiv

155. https://arxiv.org/abs/1301.3923

156. https://en.wikipedia.org/wiki/Bibcode

157. https://ui.adsabs.harvard.edu/abs/2013EPJD...67...57L

158. https://en.wikipedia.org/wiki/Digital_object_identifier

159. https://doi.org/10.1140%2Fepjd%2Fe2013-30577-8

160. https://en.wikipedia.org/wiki/International_Standard_Serial_Number

161. https://www.worldcat.org/issn/1434-6060

Loudon, R.[162] (2000). *The Quantum Theory of Light*[163] (3rd ed.). Oxford: Oxford University Press. *ISBN*[164] 978-0198501770[165]. *LCCN*[166] 2001265846[167]. *OCLC*[168] 44602993[169].

Mignani, R. P.; Testa, V.; González Caniulef, D.; Taverna, R.; Turolla, R.; Zane, S.; Wu, K. (2017). *"Evidence for vacuum birefringence from the first optical-polarimetry measurement of the isolated neutron star RX J1856.5–3754"*[170] *(PDF)*. Monthly Notices of the Royal Astronomical Society. 465 (1): 492–500. *arXiv*[171]:*1610.08323*[172]. *Bibcode*[173]:*2017MNRAS.465..492M*[174]. *doi*[175]:*10.1093/mnras/stw2798*[176]. *ISSN*[177] 0035-8711[178].

162. https://en.wikipedia.org/wiki/Rodney_Loudon

163. https://books.google.com/

     books?id=AEkfajgqldoC&printsec=frontcover#v_43ec3e5dee6e706af7766fffea512721_onepage_6cff

     047854f19ac2aa52aac51bf3af4a_q_6cff047854f19ac2aa52aac51bf3af4a_f_43ec3e5dee6e706af77

     66fffea512721_false

164. https://en.wikipedia.org/wiki/International_Standard_Book_Number

165. https://en.wikipedia.org/wiki/Special:BookSources/978-0198501770

166. https://en.wikipedia.org/wiki/Library_of_Congress_Control_Number

167. https://lccn.loc.gov/2001265846

168. https://en.wikipedia.org/wiki/OCLC

169. https://www.worldcat.org/oclc/44602993

170. https://www.eso.org/public/archives/releases/sciencepapers/eso1641/eso1641a.pdf

171. https://en.wikipedia.org/wiki/ArXiv

172. https://arxiv.org/abs/1610.08323

173. https://en.wikipedia.org/wiki/Bibcode

174. https://ui.adsabs.harvard.edu/abs/2017MNRAS.465..492M

175. https://en.wikipedia.org/wiki/Digital_object_identifier

176. https://doi.org/10.1093%2Fmnras%2Fstw2798

177. https://en.wikipedia.org/wiki/International_Standard_Serial_Number

178. https://www.worldcat.org/issn/0035-8711

Milonni, P. W.[179] (1994). *The Quantum Vacuum: An Introduction to Quantum Electrodynamics*[180]. Boston: Academic Press. *ISBN*[181] 978-0124980808[182]. *LCCN*[183] 93029780[184]. *OCLC*[185] 422797902[186].

Milonni, P. W.[187] (2009). *"Zero-Point Energy"*[188]. In Greenberger, Daniel; Hentschel, Klaus; Weinert, Friedel (eds.). Compendium of Quantum Physics: Concepts, Experiments, History and Philosophy. In D. Greenberger, K. Hentschel and F. Wienert (Eds.), Compendium of Quantum Physics Concepts, Experiments, History and Philosophy (Pp.). Berlin, Heidelberg: Springer. pp. 864–866. *arXiv*[189]:*0811.2516*[190]. *doi*[191]:*10.1007/978-3-540-70626-7*[192]. *ISBN*[193] *9783540706229*[194]. *LCCN*[195] *2008942038*[196]. *OCLC*[197] *297803628*[198].

179. https://en.wikipedia.org/wiki/Peter_W._Milonni

180. https://en.wikipedia.org/wiki/
    The_Quantum_Vacuum:_An_Introduction_to_Quantum_Electrodynamics

181. https://en.wikipedia.org/wiki/International_Standard_Book_Number

182. https://en.wikipedia.org/wiki/Special:BookSources/978-0124980808

183. https://en.wikipedia.org/wiki/Library_of_Congress_Control_Number

184. https://lccn.loc.gov/93029780

185. https://en.wikipedia.org/wiki/OCLC

186. https://www.worldcat.org/oclc/422797902

187. https://en.wikipedia.org/wiki/Peter_W._Milonni

188. https://books.google.com/books?id=ekyAV8VtfuYC&pg=PP864

189. https://en.wikipedia.org/wiki/ArXiv

190. https://arxiv.org/abs/0811.2516

191. https://en.wikipedia.org/wiki/Digital_object_identifier

192. https://doi.org/10.1007%2F978-3-540-70626-7

193. https://en.wikipedia.org/wiki/International_Standard_Book_Number

194. https://en.wikipedia.org/wiki/Special:BookSources/9783540706229

195. https://en.wikipedia.org/wiki/Library_of_Congress_Control_Number

196. https://lccn.loc.gov/2008942038

197. https://en.wikipedia.org/wiki/OCLC

198. https://www.worldcat.org/oclc/297803628

Peebles, P. J. E.[199]; Ratra, Bharat[200] (2003). "The Cosmological Constant and Dark Energy". Reviews of Modern Physics. 75 (2): 559–606. *arXiv[201]:astro-ph/0207347[202].   Bibcode[203]:2003RvMP...75..559P[204]. doi[205]:10.1103/RevModPhys.75.559[206]. ISSN[207] 0034-6861[208].*

Power, E. A.[209] (1964). Introductory Quantum Electrodynamics. London: Longmans. *LCCN[210] 65020006[211]. OCLC[212] 490279969[213].*

Rafelski, J.[214]; Muller, B. (1985). *Structured Vacuum: Thinking About Nothing[215] (PDF).* H. Deutsch: Thun. *ISBN[216] 978-3871448898[217]. LCCN[218] 86175968[219]. OCLC[220] 946050522[221].*

---

199. https://en.wikipedia.org/wiki/Jim_Peebles

200. https://en.wikipedia.org/wiki/Bharat_Ratra

201. https://en.wikipedia.org/wiki/ArXiv

202. https://arxiv.org/abs/astro-ph/0207347

203. https://en.wikipedia.org/wiki/Bibcode

204. https://ui.adsabs.harvard.edu/abs/2003RvMP...75..559P

205. https://en.wikipedia.org/wiki/Digital_object_identifier

206. https://doi.org/10.1103%2FRevModPhys.75.559

207. https://en.wikipedia.org/wiki/International_Standard_Serial_Number

208. https://www.worldcat.org/issn/0034-6861

209. https://en.wikipedia.org/wiki/Edwin_Power

210. https://en.wikipedia.org/wiki/Library_of_Congress_Control_Number

211. https://lccn.loc.gov/65020006

212. https://en.wikipedia.org/wiki/OCLC

213. https://www.worldcat.org/oclc/490279969

214. https://en.wikipedia.org/wiki/Johann_Rafelski

215. http://www.physics.arizona.edu/~rafelski/Books/StructVacuumE.pdf

216. https://en.wikipedia.org/wiki/International_Standard_Book_Number

217. https://en.wikipedia.org/wiki/Special:BookSources/978-3871448898

218. https://en.wikipedia.org/wiki/Library_of_Congress_Control_Number

219. https://lccn.loc.gov/86175968

220. https://en.wikipedia.org/wiki/OCLC

Rees, Martin[222], ed. (2012). Universe[223]. New York: DK Pub. ISBN[224]
978-0-7566-9841-6[225]. LCCN[226] 2011277855[227]. OCLC[228]
851193468[229].

Riek, C.; Seletskiy, D. V.; Moskalenko, A. S.; Schmidt, J. F.; Krauspe,
P.; Eckart, S.; Eggert, S.; Burkard, G.; Leitenstorfer, A. (2015). "Direct
Sampling of Electric-Field Vacuum Fluctuations"[230] (PDF). Science. 350
(6259): 420–423. Bibcode[231]:2015Sci...350..420R[232]. doi[233]:10.1126/
science.aac9788[234]. ISSN[235] 0036-8075[236]. PMID[237] 26429882[238].

221. https://www.worldcat.org/oclc/946050522

222. https://en.wikipedia.org/wiki/Martin_Rees

223. https://books.google.com/

   books?id=IvDZqZz0Q-QC&printsec=frontcover#v_43ec3e5dee6e706af7766fffea512721_onepage_

   6cff047854f19ac2aa52aac51bf3af4a_q_6cff047854f19ac2aa52aac51bf3af4a_f_43ec3e5dee6e706af

   7766fffea512721_false

224. https://en.wikipedia.org/wiki/International_Standard_Book_Number

225. https://en.wikipedia.org/wiki/Special:BookSources/978-0-7566-9841-6

226. https://en.wikipedia.org/wiki/Library_of_Congress_Control_Number

227. https://lccn.loc.gov/2011277855

228. https://en.wikipedia.org/wiki/OCLC

229. https://www.worldcat.org/oclc/851193468

230. http://kops.uni-konstanz.de/bitstream/handle/123456789/31877/

   Leitensdorfer_0-301750.pdf?sequence=1&isAllowed=y

231. https://en.wikipedia.org/wiki/Bibcode

232. https://ui.adsabs.harvard.edu/abs/2015Sci...350..420R

233. https://en.wikipedia.org/wiki/Digital_object_identifier

234. https://doi.org/10.1126%2Fscience.aac9788

235. https://en.wikipedia.org/wiki/International_Standard_Serial_Number

236. https://www.worldcat.org/issn/0036-8075

237. https://en.wikipedia.org/wiki/PubMed_Identifier

238. https://pubmed.ncbi.nlm.nih.gov/26429882

Rugh, S. E.; Zinkernagel, H. (2002). "The Quantum Vacuum and the Cosmological Constant Problem". Studies in History and Philosophy of Science Part B: Studies in History and Philosophy of Modern Physics. 33 (4): 663–705. *arXiv*[239]:*hep-th/0012253*[240]. *Bibcode*[241]:*2002SHPMP..33..663R*[242]. *doi*[243]:*10.1016/ S1355-2198(02)00033-3*[244]. *ISSN*[245] *1355-2198*[246].

Schwinger, J.[247] (1998a). Particles, Sources, and Fields: Volume I. Reading, Massachusetts: Advanced Book Program, Perseus Books. *ISBN*[248] *978-0-7382-0053-8*[249]. *LCCN*[250] *98087896*[251]. *OCLC*[252] *40544377*[253].

Schwinger, J.[254] (1998b). Particles, Sources, and Fields: Volume II. Reading, Massachusetts: Advanced Book Program, Perseus Books.

---

239. https://en.wikipedia.org/wiki/ArXiv

240. https://arxiv.org/abs/hep-th/0012253

241. https://en.wikipedia.org/wiki/Bibcode

242. https://ui.adsabs.harvard.edu/abs/2002SHPMP..33..663R

243. https://en.wikipedia.org/wiki/Digital_object_identifier

244. https://doi.org/10.1016%2FS1355-2198%2802%2900033-3

245. https://en.wikipedia.org/wiki/International_Standard_Serial_Number

246. https://www.worldcat.org/issn/1355-2198

247. https://en.wikipedia.org/wiki/Julian_Schwinger

248. https://en.wikipedia.org/wiki/International_Standard_Book_Number

249. https://en.wikipedia.org/wiki/Special:BookSources/978-0-7382-0053-8

250. https://en.wikipedia.org/wiki/Library_of_Congress_Control_Number

251. https://lccn.loc.gov/98087896

252. https://en.wikipedia.org/wiki/OCLC

253. https://www.worldcat.org/oclc/40544377

254. https://en.wikipedia.org/wiki/Julian_Schwinger

ISBN[255]   *978-0-7382-0054-5*[256]*.   LCCN*[257]   *98087896*[258]*.   OCLC*[259]
40544377[260]*.

Schwinger, J.[261] (1998c). Particles, Sources, and Fields: Volume III.
Reading, Massachusetts: Advanced Book Program, Perseus Books.
ISBN[262]   *978-0-7382-0055-2*[263]*.   LCCN*[264]   *98087896*[265]*.   OCLC*[266]
40544377[267]*.

Sciama, D. W.[268] (1991). "The Physical Significance of the Vacuum State
of a Quantum Field". In *Saunders, Simon*[269]; *Brown, Harvey R.*[270] (eds.).
The Philosophy of Vacuum. Oxford: Oxford University Press. *ISBN*[271]
978-0198244493[272]*. LCCN*[273] *90048906*[274]*. OCLC*[275] *774073198*[276]*.

255. https://en.wikipedia.org/wiki/International_Standard_Book_Number

256. https://en.wikipedia.org/wiki/Special:BookSources/978-0-7382-0054-5

257. https://en.wikipedia.org/wiki/Library_of_Congress_Control_Number

258. https://lccn.loc.gov/98087896

259. https://en.wikipedia.org/wiki/OCLC

260. https://www.worldcat.org/oclc/40544377

261. https://en.wikipedia.org/wiki/Julian_Schwinger

262. https://en.wikipedia.org/wiki/International_Standard_Book_Number

263. https://en.wikipedia.org/wiki/Special:BookSources/978-0-7382-0055-2

264. https://en.wikipedia.org/wiki/Library_of_Congress_Control_Number

265. https://lccn.loc.gov/98087896

266. https://en.wikipedia.org/wiki/OCLC

267. https://www.worldcat.org/oclc/40544377

268. https://en.wikipedia.org/wiki/Dennis_W._Sciama

269. https://en.wikipedia.org/wiki/Simon_Saunders

270. https://en.wikipedia.org/wiki/Harvey_Brown_(philosopher)

271. https://en.wikipedia.org/wiki/International_Standard_Book_Number

272. https://en.wikipedia.org/wiki/Special:BookSources/978-0198244493

273. https://en.wikipedia.org/wiki/Library_of_Congress_Control_Number

274. https://lccn.loc.gov/90048906

275. https://en.wikipedia.org/wiki/OCLC

Saunders, Simon[277]; Brown, Harvey R.[278], eds. (1991). The Philosophy of Vacuum. Oxford: Oxford University Press. *ISBN[279] 978-0198244493[280]. LCCN[281] 90048906[282]. OCLC[283] 774073198[284].*

Urban, M.; Couchot, F.; Sarazin, X.; Djannati-Atai, A. (2013). "The Quantum Vacuum as the Origin of the Speed of Light". The European Physical Journal D. 67 (3): 58. *arXiv[285]:1302.6165[286]. Bibcode[287]:2013EPJD...67...58U[288]. doi[289]:10.1140/epjd/ e2013-30578-7[290]. ISSN[291] 1434-6060[292].*

Weinberg, S.[293] (1989). "The Cosmological Constant Problem"[294] (PDF). Reviews of Modern Physics. 61 (1): 1–23.

---

276. https://www.worldcat.org/oclc/774073198

277. https://en.wikipedia.org/wiki/Simon_Saunders

278. https://en.wikipedia.org/wiki/Harvey_Brown_(philosopher)

279. https://en.wikipedia.org/wiki/International_Standard_Book_Number

280. https://en.wikipedia.org/wiki/Special:BookSources/978-0198244493

281. https://en.wikipedia.org/wiki/Library_of_Congress_Control_Number

282. https://lccn.loc.gov/90048906

283. https://en.wikipedia.org/wiki/OCLC

284. https://www.worldcat.org/oclc/774073198

285. https://en.wikipedia.org/wiki/ArXiv

286. https://arxiv.org/abs/1302.6165

287. https://en.wikipedia.org/wiki/Bibcode

288. https://ui.adsabs.harvard.edu/abs/2013EPJD...67...58U

289. https://en.wikipedia.org/wiki/Digital_object_identifier

290. https://doi.org/10.1140%2Fepjd%2Fe2013-30578-7

291. https://en.wikipedia.org/wiki/International_Standard_Serial_Number

292. https://www.worldcat.org/issn/1434-6060

293. https://en.wikipedia.org/wiki/Steven_Weinberg

294. https://repositories.lib.utexas.edu/bitstream/2152/61094/1/Weinberg_1989.pdf

Bibcode[295]:1989RvMP...61....1W[296]. doi[297]:10.1103/RevModPhys.61.1[298]. hdl[299]:2152/61094[300]. ISSN[301] 0034-6861[302].

Weinberg, S.[303] (2015). Lectures on Quantum Mechanics (2nd ed.). Cambridge: Cambridge University Press. *ISBN[304] 978-1-107-11166-0[305]. LCCN[306] 2015021123[307]. OCLC[308] 910664598[309].*

Weisskopf, V. (1936). *"Über die Elektrodynamik des Vakuums auf Grund des Quanten-Theorie des Elektrons'[310] (PDF).* Kongelige Danske Videnskabernes Selskab, Mathematisk-fysiske Meddelelse. 24 (6): 3–39.

Wilson, C. M.; Johansson, G.; Pourkabirian, A.; Simoen, M.; Johansson, J. R.; Duty, T.; Nori, F.; Delsing, P. (2011). "Observation of the Dynamical Casimir Effect in a Superconducting Circuit". Nature. 479 (7373): 376–379. *arXiv[311]:1105.4714[312].*

---

295. https://en.wikipedia.org/wiki/Bibcode

296. https://ui.adsabs.harvard.edu/abs/1989RvMP...61....1W

297. https://en.wikipedia.org/wiki/Digital_object_identifier

298. https://doi.org/10.1103%2FRevModPhys.61.1

299. https://en.wikipedia.org/wiki/Handle_System

300. https://hdl.handle.net/2152%2F61094

301. https://en.wikipedia.org/wiki/International_Standard_Serial_Number

302. https://www.worldcat.org/issn/0034-6861

303. https://en.wikipedia.org/wiki/Steven_Weinberg

304. https://en.wikipedia.org/wiki/International_Standard_Book_Number

305. https://en.wikipedia.org/wiki/Special:BookSources/978-1-107-11166-0

306. https://en.wikipedia.org/wiki/Library_of_Congress_Control_Number

307. https://lccn.loc.gov/2015021123

308. https://en.wikipedia.org/wiki/OCLC

309. https://www.worldcat.org/oclc/910664598

310. http://neo-classical-physics.info/uploads/3/0/6/5/3065888/weisskopf_-_electrodynamics.pdf

311. https://en.wikipedia.org/wiki/arXiv

Bibcode[313]:*2011Natur.479..376W*[314].   *doi*[315]:*10.1038/nature10561*[316]. *ISSN*[317] *0028-0836*[318]. *PMID*[319] *22094697*[320].

Balungi Francis, (2010) "A hypothetical investigation into the realm of the microscopic and macroscopic universes beyond the standard model" general physics arXiv:1002.2287v1[321] [physics.gen-ph]

Hawking, Stephen[322] (1975). "Particle Creation by Black Holes"[323]. Commun.   Math.   Phys.[324]   43   (3):   199–220. Bibcode[325]:1975CMaPh..43..199H[326].

Hawking,   S.   W.[327]   (1974).   "Black   hole   explosions?". Nature.248(5443):30–31.

Bibcode[328]:1974Natur.248...30H[329].doi[330]:10.1038/248030a0[331].

312. https://arxiv.org/abs/1105.4714

313. https://en.wikipedia.org/wiki/Bibcode

314. https://ui.adsabs.harvard.edu/abs/2011Natur.479..376W

315. https://en.wikipedia.org/wiki/Digital_object_identifier

316. https://doi.org/10.1038%2Fnature10561

317. https://en.wikipedia.org/wiki/International_Standard_Serial_Number

318. https://www.worldcat.org/issn/0028-0836

319. https://en.wikipedia.org/wiki/PubMed_Identifier

320. https://pubmed.ncbi.nlm.nih.gov/22094697

321. https://arxiv.org/abs/1002.2287v1

322. https://en.wikipedia.org/wiki/Stephen_Hawking

323. http://www.springerlink.com/content/c4553033029k5wk6/

324. https://en.wikipedia.org/wiki/Commun._Math._Phys.

325. https://en.wikipedia.org/wiki/Bibcode

326. http://adsabs.harvard.edu/abs/1975CMaPh..43..199H

327. https://en.wikipedia.org/wiki/Stephen_Hawking

328. https://en.wikipedia.org/wiki/Bibcode

329. http://adsabs.harvard.edu/abs/1974Natur.248...30H

330. https://en.wikipedia.org/wiki/Digital_object_identifier

Carlo Rovelli (2003) "Quantum Gravity" Draft of the Book Pdf

Some few texts used are from Wikipedia https://creativecommons.org/
licenses/by-sa/3.0/

D. N. Page, Phys. Rev. D 13, 198 (1976).

C. Gao and Y.Lu, Pulsations of a black hole in LQG (2012)
arXiv:1706.08009v3

A.H. Chamseddine and V.Mukhanov, Non singular black hole (2016)
arXiv 1612.05861v1

M.Bojowald and G.M.Paily, A no-singularity scenario in LQG (2012)
arXiv: 1206.5765v1

P.Singh, class.Quant.Grav,26,125005(2009), arXiv:0901.2750

P.Singh and F.Vidotto, Phys.Rev, D83,064027(2011) arXiv:1012.1307

C.Rovelli   and   F.Vidotto,   Phy.   Rev,111(9)   091303(2013)
arXiv:1307.3228v2

M.Bojowald, Initial conditions for a universe, Gravity Research
Foundation (2003)

A.Ashtekar, Singularity Resolution in Loop Quantum Cosmology
(2008) arXiv:0812.4703v1

J.Brunneumann and T.Thiemann, On singularity avoidance in Loop
Quantum Gravity (2005) arXiv:0505032v1

L.Modesto, Disappearence of the Black hole singularity in Quantum
gravity (2004) arXiv:0407097v2

Mikhailov, A.A. (1959).Mon. Not. Roy. Astron. Soc.,119, 593.

P. Merat etal.(1974). Astron & Astrophys 32, 471-475

Trempler, R.J. (1956).Helv. Phys. Acta, Suppl.,IV, 106.

Trempler, R.J. (1932). " The deflection of light in the sun's gravitational field "Astronomical society of the pacific 167

Einstein, A. (1916).Ann. d. Phys.,49, 769; (1923).The Principle of Relativity, (translators Perret, W. and Jeffery, G.B.), (Dover Publications, Inc., New York), pp. 109–164.

Von Klüber, H. (1960). InVistas in Astronomy, Vol. 3, pp. 47–77.

K. Hentschel (1992). Erwin Finlay Freundlich and testing Einstein theory of relativity, Communicated by J.D. North

Muhleman, D.O., Ekers, R.D. and Fomalont, E.B. (1970).Phys. Rev. Lett.,24, 1377

Mikhailov, A.A. (1956).Astron. Zh.,33, 912.

Dyson, F.W., Eddington, A.S. and Davidson, C. (1920).Phil. Trans. Roy. sog., A220, 291

Chant, C.A. and Young, R.K. (1924).Publ. Dom. Astron. Obs.,2, 275.

Campbell, W.W. and Trumbler, R.J. (1923).Lick Obs. Bull.,11, 41.

Freundlich, E.F., von Klüber, H. and von Brunn, A. (1931).Abhandl. Preuss. Akad. Wiss. Berlin, Phys. Math. Kl., No.l;Z. Astrophys.,3, 171

Mikhailov, A.A. (1949).Expeditions to Observe the Total Solar Eclipse of 21 September, 1941, (report), (ed. Fesenkov, V.G.), (Publications of the Academy of Sciences, U.S.S.R.), pp. 337–351.

S.P. Martin, in Perspectives on Supersymmetry , edited by G.L. Kane (World Scientific, Singapore, 1998) pp. 1–98; and a longer archive version in hep-ph/9709356; I.J.R. Aitchison, hep-ph/0505105.

Mara Beller, Quantum Dialogue: The Making of a Revolution. University of Chicago Press, Chicago, 2001.

Morrison, Philp: "The Neutrino, scientific American, Vol 194,no.1 (1956),pp.58-68.

R. Haag, J. T. Lopuszanski and M. Sohnius, Nucl. Phys. B88, 257 (1975) S.R. Coleman and J. Mandula, Phys.Rev. 159 (1967) 1251.

H.P. Nilles, Phys. Reports 110, 1 (1984).

P. Nath, R. Arnowitt, and A.H. Chamseddine, Applied N = 1 Supergravity (World Scientific, Singapore, 1984).

S.P. Martin, in Perspectives on Supersymmetry , edited by G.L. Kane (World Scientific, Singapore, 1998) pp. 1–98; and a longer archive version in hep-ph/9709356; I.J.R. Aitchison, hep-ph/0505105.

S. Weinberg, The Quantum Theory of Fields, VolumeIII: Supersymmetry (Cambridge University Press, Cambridge,UK, 2000).

E. Witten, Nucl. Phys. B188, 513 (1981).

S. Dimopoulos and H. Georgi, Nucl. Phys. B193, 150(1981).

N. Sakai, Z. Phys. C11, 153 (1981);R.K. Kaul, Phys. Lett. 109B, 19 (1982).

L. Susskind, Phys. Reports 104, 181 (1984).

L. Girardello and M. Grisaru, Nucl. Phys. B194, 65(1982); L.J. Hall and L. Randall,

Phys. Rev. Lett. 65, 2939(1990);I. Jack and D.R.T. Jones, Phys. Lett. B457, 101 (1999).

For a review, see N. Polonsky, Supersymmetry: Structureand phenomena. Extensions of the standard model, Lect.Notes Phys. M68, 1 (2001).

G. Bertone, D. Hooper and J. Silk, Phys. Reports 405, 279 (2005).

G. Jungman, M. Kamionkowski, and K. Griest, Phys. Reports 267, 195 (1996).

V. Agrawal, S.M. Barr, J.F. Donoghue and D. Seckel,Phys. Rev. D57, 5480 (1998).

N. Arkani-Hamed and S. Dimopoulos, JHEP 0506, 073(2005); G.F. Giudice and A. Romanino, Nucl. Phys. B699, 65(2004) [erratum: B706, 65 (2005)]. July 27, 2006 11:28

en.wikipedia.org/wiki/Supersymmetry - 52k - Cached[332] - Similar pages[333]

en.wikipedia.org/wiki/Grand_unification_theory - 39k - Cached[334] - Similar pages[335]

In cosmology, the Planck epoch (or Planck era), named after Max Planck, is the earliest period of time in the history of the universe, en.wikipedia.org/wiki/**Planck_epoch** - 23k - Cached[336] - Similar pages

---

332. http://64.233.169.104/search?q=cache:ZBSZWNLrdxEJ:en.wikipedia.org/wiki/

  Supersymmetry+supersymmetry&hl=en&ct=clnk&cd=1&gl=ug

333. http://www.google.co.ug/search?hl=en&q=related:en.wikipedia.org/wiki/Supersymmetry

334. http://64.233.169.104/search?q=cache:Le3KZW7QnUYJ:en.wikipedia.org/wiki/

  Grand_unification_theory+grand+unification&hl=en&ct=clnk&cd=1&gl=ug

335. http://www.google.co.ug/search?hl=en&q=related:en.wikipedia.org/wiki/

  Grand_unification_theory

L. Shapiro and J. Sol`a, Phys. Lett. B 530, 10 (2002);

E. V.Gorbar and I. L. Shapiro, JHEP 02, 021 (2003); A. M. Pelinson,

L. Shapiro, and F. I. Takakura, Nucl. Phys. B 648, 417 (2003).

A. Starobinsky, Phys. Lett. B 91, 99 (1980).

G. F. R. Ellis, J. Murugan, and C. G. Tsagas, Class. Quant. Grav.21, 233 (2004).

H. V. Peiris et al., Astrophys. J. Suppl. 148, 213 (2003).

D. N. Spergel et al., astro-ph/0603449.

Vilenkin, Phys. Rev. D 32, 2511 (1985).

A. Starobinsky, Pis'ma Astron. Zh 9, 579 (1983).

A.H. Guth, Phys. Rev. D23, 347 (1981).

A.D. Linde, Phys. Lett. B108, 389 (1982); A. Albrecht, P.J.

Steinhardt, Phys.Rev. Lett. 48, 1220 (1982).

A.D. Linde, Phys Lett. B129, 177 (1983).

N. Makino, M. Sasaki, Prog. Theor. Phys. 86, 103 (1991);

D. Kaiser, Phys. Rev.D52, 4295 (1995).

H. Goldberg, Phys. Rev. Lett. 50, 1419 (1983).

E. Kolb and M. Turner, The Early Universe (Addison-Wesley, Reading, MA,1990).

---

336. http://64.233.169.104/search?q=cache:d5zWrem4T08J:en.wikipedia.org/wiki/
    Planck_epoch+planck+epoch&hl=en&ct=clnk&cd=1

W. Garretson and E. Carlson, Phys. Lett. B 315, 232(1993); H. Goldberg, hep-ph/0003197.

Eddington, A. S., The Internal Constitution of the Stars (Cambridge University Press, England,1926), p. 16

Duncan R .C. & Thompson C., Ap.J.392, L 9 (1992).

Thompson , C, Duncan , R .C ., Woods , P., Kouveliotou , C ., Finger , M.H. & van Parad ij s , J .,ApJ, submitted , astro-ph /9908086, (2000).

Schwinger , J .,Phys. Rev.73, 416L (1948)

Carlip, S.: Quantum gravity: a progress report. Rept. Prog. Phys. 64, 885 (2001).arXiv:gr-qc/0108040

Kerr,R.P.: Gravitational field of a spinning mass as an example of algebraically special metrics.

Phys. Rev. Lett. 11, 237–238 (1963)

Bekenstein, J.: Black holes and the second law. Lett. Nuovo Cim. 4, 737–740 (1972)

Bardeen, J.M., Carter, B., Hawking, S.: The four laws of black hole mechanics. Commun.

Math. Phys. 31, 161–170 (1973)

Tolman, R.: Relativity, Thermodynamics, and Cosmology. Dover Books on Physics Series.

Dover Publications, New York (1987)

Oppenheimer, J., Volkoff, G.: On massive neutron cores. Phys. Rev. 55, 374–381 (1939)

Tolman, R.C.: Static solutions of einstein's field equations for spheres of fluid, Phys. Rev. 55,364–373 (1939)

Zel'dovich Y.B.: Zh. Eksp. Teoret. Fiz.41, 1609 (1961)

Bondi, H.: Massive spheres in general relativity. Proc. Roy. Soc. Lond. A281, 303–317 (1964)

Sorkin, R.D., Wald, R.M., Zhang, Z.J.: Entropy of selfgravitating radiation. Gen. Rel. Grav. 1127–1146 (1981)

Newman, E.T., Couch, R., Chinnapared, K., Exton, A., Prakash, A., et al.: Metric of a rotating,charged mass. J. Math. Phys. 6, 918–919 (1965)

Ginzburg, V., Ozernoi, L.: Sov. Phys. JETP 20, 689 (1965)

Doroshkevich, A., Zel'dovich, Y., Novikov I.: Gravitational collapse of nonsymmetric and rotating masses, JETP 49 (1965)

Israel, W.: Event horizons in static vacuum space-times. Phys. Rev. 164, 1776–1779 (1967)

Israel,W.: Event horizons in static electrovac space-times. Commun. Math. Phys. 8, 245–260 (1968)

Loop quantum gravity does provide such a prediction [363, 364], and it disagrees with the semiclassical

Carter, B.: Axisymmetric black hole has only two degrees of freedom. Phys. Rev. Lett. 26, 331–333(1971)

Penrose, R.: Gravitational collapse: the role of general relativity. Riv. Nuovo Cim. 1, 252–276 (1969)

Christodoulou, D.: Reversible and irreversible transformations in black hole physics. Phys. Rev. Lett. 25, 1596–1597 (1970)

Christodoulou, D., Ruffini, R.: Reversible transformations of a charged black hole. Phys. Rev. D4, 3552–3555 (1971)

Hawking, S.: Particle creation by black holes. Commun. Math. Phys. 43, 199–220 (1975)

Klein, O.: Die reflexion von elektronen an einem potential sprung nach der relativistischen dynamik von dirac. Z. Phys. 53, 157 (1929)

Wald, R.M.: General Relativity. The University of Chicago Press, Chicago (1984)

Hawking, S.W.: Black hole explosions. Nature 248, 30–31 (1974)

Hawking, S., Ellis, G.: The large scale structure of space-time. Cambridge University Press, Cambridge (1973)

Carter, B.: Black hole equilibrium states, In Black Holes—Les astres occlus. Gordon and Breach Science Publishers, (1973)

Hawking, S.W.: Gravitational radiation from colliding black holes. Phys. Rev. Lett. 26, 1344– 1346 (1971)

Hawking, S.: Black holes in general relativity. Commun. Math. Phys. 25, 152–166 (1972)

Bekenstein, J.: Extraction of energy and charge from a black hole. Phys. Rev. D7, 949–953 (1973)

Bekenstein, J.D.: Black holes and entropy. Phys. Rev. D7, 2333–2346 (1973)

Hawking, S.: Quantum gravity and path integrals. Phys. Rev. D18, 1747–1753 (1978)

Gross, D.J., Perry, M.J., Yaffe, L.G.: Instability of flat space at finite temperature. Phys. Rev. D25, 330–355 (1982)

Unruh, W.G., Wald, R.M.: What happens when an accelerating observer detects a rindler particle. Phys. Rev. D29, 1047–1056 (1984)

Bekenstein, J.D.: Auniversal upper bound on the entropy to energy ratio for bounded systems. Phys. Rev. D23, 287 (1981)

Unruh,W.,Wald, R.M.: Acceleration radiation and generalized second law of thermodynamics. Phys. Rev. D25, 942–958 (1982)

Unruh, W., Wald, R.M.: Entropy bounds, acceleration radiation, and the generalized second law. Phys. Rev. D27, 2271–2276 (1983)

Image : MPI for gravitational physics / W.Benger-zib

Tomilin,K.A., (1999). "Natural Systems Of Units: To The Centenary Aniniversary Of The Planck Systems", 287-296

Sivaram, C. (2007). "What Is Special About the Planck Mass"? arXiv:0707.0058v1

H. Georgi and S.L. Glahow. (1974) "Unity Of All Elementary-Particle Forces". Phys. Rev. Letters 32, 438

Luigi Maxmilian Caligiuri, Amrit Sorli. Gravity Originates from Variable Energy Density of Quantum Vacuum. American Journal of Modern Physics. Vol. 3, No. 3, 2014, pp. 118-128. doi: 10.11648/ j.ajmp.20140303.11

Philip J. Tattersall,(2018) Quantum Vacuum Energy and the Emergence of Gravity. doi:10.5539/apr.v10n2p1

H. E. Puthoff (1989) Gravity as a zero-point-fluctuation force PHYSICAL REVIEW A VOLUME 39, NUMBER 5

Balungi Francis (2018) "Quantum Gravity in a Nutshell1" Book.

E.Verlinde (2016) Emergent Gravity and the Dark Universe, arXiv:1611.02269v2[hep-th]

S.Hossenfelder (2018) The Redshift-Dependence of Radial Acceleration: Modified gravity versus particle dark matter, arXiv:1803.08683v1[gr-qc]

Robert J. Scherrer (2004) Purely kinetic k-essence as unified dark matter, arXiv:astro-ph/0402316v3

J.S.Farnes (2018), Aunifying theory of dark energy and dark matter: Negative masses and matter creation within a modified ΛCDM framework, arXiv:1712.07962v2[physics.gen-ph]

Gustav M Obermair (2013), Primordial Planck mass black holes (PPMBHs) as candidates for dark matter? J. Phys:conf.Ser.442012066

V.Cooray etal...(2017), An alternative approach to estimate the vacuum energy density of free space, doi:10.20944/preprints201707.0048.v1

M.Milgrom, (1983) A modification of the Newtonian dynamics: Implications for galaxies, Astrophys.J.270, 371.

Acknowledgments

This book would never have been completed without the patience and dedication of my wife, Wanyana Ritah. She performed the wonderful and difficult task of editing major parts of the book and helped in researching many details necessary to complete it.

I wish to thank several colleagues for their help and extensive comments on the manuscript. I particularly thank a total of 200 online physics friends and SUSP science foundation members, for a careful reading of the manuscript. Many graduate students have contributed over the years to developing my Quantum theory of gravity.

I also wish to thank my editors, at SUSP science Foundation and Bill stone Services for their enthusiasm and support. Finally, I thank our family for their patience, love, and support during the four years of working on this book.

About the Author

Balungi Francis was born in Kampala, Uganda, to a single poor mother, grew up in Kawempe, and later joined Makerere Universty in 2006, graduating with a Bachelor Science degree in Land Surveying in 2010. For four years he taught in Kampala City high schools, majoring in the fields of Gravitation and Quantum Physics. His first book, "Mathematical Foundation of the Quantum theory of Gravity," won the Young Kampala Innovative Prize and was mentioned in the African Next Einstein Book Prize (ANE).

He has spent over 15years researching and discovering connections in physics, mathematics, geometry, cosmology, quantum mechanics, gravity, in addition to astrophysics, unified physics and geographical information systems . These studies led to his groundbreaking theories, published papers, books and patented inventions in the science of Quantum Gravity, which have received worldwide recognition.

From these discoveries, Balungi founded the SUSP (Solutions to the Unsolved Scientific Problems) Project Foundation in 2004 - now known as the SUSP Science Foundation. As its current Director of Research, Balungi leads physicists, mathematicians and engineers in exploring Quantum Gravity principles and their implications in our world today and for future generations.

Balungi launched the Visionary School of Quantum Gravity in 2016 in order to bring the learning and community further together. It's the first and only Quantum Gravity physics program of its kind, educating thousands of students from over 80 countries.

The book "Quantum Gravity in a Nutshell1", a most recommend book in quantum gravity research, was produced based on Balungi's discoveries and their potential for generations to come. Balungi is currently guiding the Foundation, speaking to audiences worldwide, and continuing his groundbreaking research.

Contact Balungi Francis at the following address:

Email: balungif@gmail.com

bfrancis@cedat.mak.ac.ug

BILL STONE SERVICES

Tel: +256703683756

+256777105605

Did you love *Unified Field Theory in a Nutshell1: The Quest for the Theory of Everything*? Then you should read *Fifty Formulas that Changed the World*[1] by Balungi Francis!

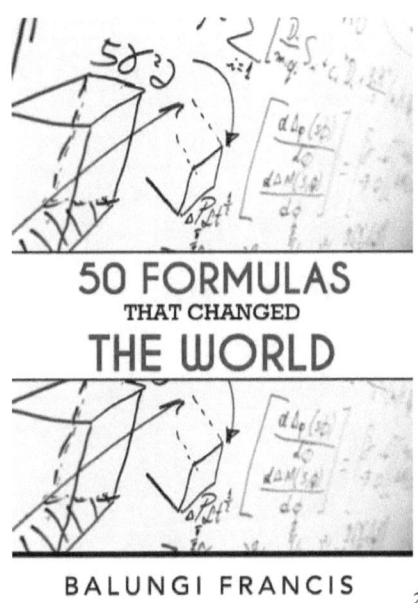

50 FORMULAS
THAT CHANGED
THE WORLD

BALUNGI FRANCIS [2]

Balungi has discovered the secret formulas behind the Kentucky Fried Chicken (KFC), Coca-Cola, Facebook, Google, the Theory of Everything, Quantum Gravity and the list is endless. Fifty of the most important equations in physics, economics, mathematics, catering, beverages are all given and expalined in his bestselling book.

LEARN THE FORMULAS BILLIONAIRES ARE USING TO OUT COMPETE YOU AND ENTER 2020 WITH SUCCESS.

LEARN FORMULAS BEHIND EVERY MULTI BILLION BUSINESS AND START YOUR OWN.

---

1. https://books2read.com/u/31GON7

2. https://books2read.com/u/31GON7

www.ingramcontent.com/pod-product-compliance
Lightning Source LLC
Chambersburg PA
CBHW022132170526
45157CB00004B/1843